# PRESERVED SHUNTERS

ANDREW COLE



AMBERLEY

*I would like to dedicate this book to Steve Blick: family man, shunter enthusiast, organiser of many trips and, above all, a true friend – RIP*

First published 2021

Amberley Publishing
The Hill, Stroud
Gloucestershire, GL5 4EP

www.amberley-books.com



ISBN 978 1 3981 0650 5 (print)
ISBN 978 1 3981 0651 2 (ebook)



British Library Cataloguing in Publication Data.
A catalogue record for this book is available from the British Library.

Typesetting by SJmagic DESIGN SERVICES, India.
Printed in the UK.

# Contents

# Introduction

Over the years British Rail have employed the services of thousands of shunt engines. Built at both their own workshops, and by many different manufacturers, they could be found working at almost every station, yard and depot all over the country.

The largest class of loco was the British Rail Class 08, with nearly a thousand examples built. But there were also small batches made by private companies hoping to win large orders from British Rail. These included Andrew Barclay, Hudswell, Ruston & Hornsby, North British, the Yorkshire Engine Company and Hunslet.

Many were built for specific areas or regions; for example, the Ruston-built Class 07 locomotives were made for use on the Southern Region around the tight curves of Southampton Docks. The Andrew Barclay-built Class 06 locomotives were used in Scotland, except for when No. 06003 entered departmental service at Reading.

Many of the locomotives were based on the 0-6-0 wheel design, but a small number were also built to the smaller 0-4-0 design.

When the movement of freight on the railways began to contract with competition from the road hauliers there became less of a need for so many shunt engines. Some of the smaller batches of locomotives managed less than ten years of service, and many classes were withdrawn prior to being given a TOPS classification number series.

Those classes that did survive had large numbers withdrawn and sold for scrap before British Rail concentrated on keeping members of the Class 03, 08 and 09 in service. This resulted in a large number of useable locomotives being made available to purchase for industrial use. A couple of industries that made full use of the locomotives available were the National Coal Board and British Steel, buying large numbers of withdrawn locomotives and putting them to good use. Smaller companies also brought locomotives for use at their depots and on their lines – from Gulf Oil to Redland Roadstone. There were large numbers put to use.

A large number of locomotives finished their second careers in industry and found themselves the subject of preservation bids. If not for the move into industry some classes may have become extinct altogether – most notably the Class 02 and 04, but also the larger Class 14.

A prolific purchaser of redundant shunters was the South Yorkshire Railway Preservation Society. Initially based at Attercliffe in Yorkshire, they moved to larger premises at Meadowhall near Sheffield. They purchased members of nearly every class of loco from industrial sites all over the country. When the society had to move out of their base in Sheffield, they created the Heritage Shunter Trust in Rowsley, Derbyshire, and moved their many locomotives to their new site in the early 1990s.

Many different preservation sites gained from the use of shunters that were brought from industry, and this book is a testament to the hard work that all the different sites have put into saving and restoring the shunters for people to enjoy today.

# Class 01

**D2953, 3 September 2017**
D2953 is seen stabled outside the Heritage Shunter Trust shed at Rowsley, Derbyshire. D2953 was one of five shunters built by Andrew Barclay, and today two survive. D2953 was withdrawn in 1966 and was sold to Thames Matex of West Thurrock, Essex. It would be preserved in 1985 by the South Yorkshire Preservation Society before moving to Rowsley.

**D2956, 8 July 2016**
D2956 is seen preserved at the East Lancashire Railway, Bury, carrying black livery. D2956 was also withdrawn in 1966 and was brought by A. King for use as a shunter at its Norwich scrapyard. It later moved to King's yard at Snailwell – King's was later brought by Meyer Newman. The loco was preserved at Bury in 1985.

# Class 02

**No. 02003 (D2853), 21 July 2019**

No. 02003 stands inside the roundhouse at Barrow Hill carrying green livery. This was one of twenty locomotives built by the Yorkshire Engine Company in 1960, and is the only preserved member that lasted long enough in BR service to receive a TOPS number. It was withdrawn in 1975 and sold to LCP Fuels LTD in Brierley Hill before it entered preservation in 1999 at the Rutland Railway Museum.

**D2854, 3 September 2017**

D2854 is seen outside the Heritage Shunter Trust depot at Rowsley. This locomotive was withdrawn in 1970 and was sold to CF Booth, Rotherham, for use as a yard shunter. It would be preserved by the South Yorkshire Railway Preservation Society, and moved to Meadowhall in 1988 before moving to Rowsley.

**D2858, 14 July 2019**
D2858 stands inside the museum building at Butterley on the Midland Railway Centre. This was another Class 02 locomotive that was withdrawn in 1970. It was sold to Hutchinson Estate & Dock Company in Widnes. It was later sold to Fisons Fertilizers, also in Widnes, before being sold for scrap to Lowton Metals. It was later moved in 1986 to the Midland Railway Centre.

**D2860, 1 September 2019**
D2860 is seen outside the diesel shed at the National Railway Museum, York. This shunter was withdrawn in 1970 and later moved to Preston Park Pullman Works, Brighton, before moving to York as part of the National Collection in 1979.

**D2866, 29 May 2016**
D2866 is seen undergoing a very thorough overhaul inside the Heritage Shunter Trust maintenance shed at Rowsley. This was withdrawn in 1970 and was brought by Arnott Young in Dalmuir for further use. It was later sold to the Brechin Railway, who later sold it on to the South Yorkshire Railway Preservation Society in 1996.

**D2867, 23 August 2009**
D2867 *Diane* is seen at Shackerstone on the Battlefield Line. This was withdrawn in 1970 and was sold to Redland Roadstone for use at Mountsorrel, later moving to Barrow-on-Soar. It is seen carrying Redland livery and came to the Battlefield Line via Meadowhall.

**D2868, 21 July 2019**
D2868 is seen inside the roundhouse at Barrow Hill, undergoing attention, with its engine covers removed. This was another shunter brought by LCP Fuels, Brierley Hill, in 1970. It would eventually be sold to the Battlefield Line in 2001, before moving to Peak Rail, Rowsley.

# Class 03

**No. 2022 (No. 03022), 30 July 2006**
No. 2022 was one of over 200 Class 03 locomotives built from 1957 onwards. They were built by British Railways at their Doncaster and Swindon workshops. No. 2022 was withdrawn in 1982 from Gateshead depot, Newcastle, and was sold directly to the Swindon & Cricklade Railway where it is seen carrying a cast GWR number plate.

**D2023, 14 May 2016**
D2023 is seen preserved at Tenterden on the Kent & East Sussex Railway. This was withdrawn in 1971 and was sold to the Tees & Hartlepool Dock Authority, before being sold to the K&ESR for preservation in 1983.

**D2024, 14 May 2016**
D2024 is seen in storage at Bodiam station on the Kent & East Sussex Railway. This loco shared a remarkable life with consecutively numbered D2023, being sold to the Tees & Hartlepool Dock Authority, before being sold with its classmate to the K&ESR in 1983.

**No. 03027 (D2027), 3 September 2017**
No. 03027 is seen waiting its turn in the restoration queue at the Heritage Shunter Trust site at Rowsley. This locomotive was withdrawn from BR service in 1976 and was brought by P. Wood, Queenborough. It was later preserved by the South Yorkshire Railway Preservation Society in 1991.

**No. 03037 (D2037), 18 August 2018**
D2037 is seen stabled among the trees at the Royal Deeside Railway, Aberdeenshire. This loco was withdrawn in 1976 and was brought by Hargreaves Industrial Services in Wakefield. It later moved to British Oak at Oxcroft Opencast Coal Disposal Point, Derbyshire, before being preserved at Meadowhall. It later moved to the Lavender Line before heading to Scotland.

**D2041, 23 September 2017**
D2041 is seen carrying BR green livery at Castle Hedingham on the Colne Valley Railway. Withdrawn in 1970, D2041 was brought by the Central Electricity Generating Board, who used it at various sites, including Richborough, Rye House and Barking. It was sold to the Colne Valley Railway in 1981.

**D2046, 18 November 2018**
D2046 is seen undergoing restoration at Plym Bridge depot at the Plym Valley Railway, Plymouth. This was withdrawn from Thornaby depot in 1971 and was sold to Gulf Oil, Waterston, before entering preservation.

**D2051, 29 August 2018**
D2051 is seen outside the carriage shed at Holt on the North Norfolk Railway, waiting its turn for restoration. This was withdrawn in 1972 and was sold to the Ford Motor Company for use at Dagenham. It was later brought for preservation, arriving at the North Norfolk Railway in 2000.

**No. 03059 (D2059), 1 October 2016**
D2059 departs from Haven Street on the Isle of Wight Steam Railway, heading for Smallbrook Junction. This was withdrawn from BR service in 1987 and was sold directly to the IOWSR.

**No. 03062 (D2062), 16 April 2016**
D2062 arrives at Bury Bolton Street on the East Lancashire Railway. D2062 was withdrawn in 1980 and was sold directly for preservation to the Dean Forest Railway. It was later sold to the East Lancashire Railway in 1997.

**No. 03063 (D2063), 20 July 2006**
D2063 is seen carrying BR blue livery at Sheringham on the North Norfolk Railway – one of two Class 03s based on the line. D2063 was withdrawn in 1987 and was sold to the Colne Valley Railway, before it was resold to the North Norfolk Railway, arriving in 2000.

**No, 03066 (D2066), 11 September 2016**
No. 03066 sits on the turntable inside the roundhouse at Barrow Hill. This was withdrawn from Gateshead depot in 1988 and was brought by Parkfield Castings, Horwich. It was brought for preservation by the South Yorkshire Railway Preservation Society in 1991 and later moved to Barrow Hill in 1999.

**No. 03073 (D2073), 21 August 1994**
No. 03073 is seen on display at Basford Hall open day, Crewe. This was a late survivor with BR, not being withdrawn until 1989 from Birkenhead depot before being sold directly to the Railway Age, Crewe, for preservation.

**No. 03078 (D2078), 8 October 2017**
D2078 stands in the yard at the Stephenson Railway Museum, Tyneside, carrying North Tyneside Railway livery. This was withdrawn in 1988 and moved directly to the museum from Gateshead.

**No. 03081 (D2081), 17 June 2017**
No. 03081 is seen at Mangapps Farm Railway Museum coupled to a match truck. This loco was withdrawn in 1980 and was sold for further industrial use in Belgium, before ending up in use at Gruppe sugar factory. It was repatriated in 2004 and carries the name *Lucie*.

**No. 03084 (D2084), 20 April 2003**
No. 03084 is seen in use as yard pilot at Darley Dale, Peak Rail. This was withdrawn in 1987 and was sold directly into preservation, initially at the Ecclesbourne Valley Railway. It later moved to Peak Rail, and then even later to the Lincolnshire Wolds Railway, and can today be found at West Coast Railways, Carnforth.

**No. 03084 (D2084), 16 April 2016**
D2084 is seen at Bury Bolton Street on the East Lancashire Railway. This was operating in top-and-tail mode with classmate D2062 during the line's small diesels weekend. It has since moved to Carnforth.

**No. 03090 (D2090), 7 October 2017**
D2090 is seen carrying BR green livery at Shildon Locomotion Museum. This loco forms part of the National Collection. It was withdrawn from BR service in 1976 and moved directly to the National Railway Museum. Note the exhaust scrubber fitted to allow the loco to work inside the museum.

**No. 03094 (D2094), 18 August 2018**
D2094 is seen stabled at Milton of Crathes station on the Royal Deeside Railway, Aberdeenshire. There are three Class 03 locomotives based on the railway. D2094 was withdrawn in 1988 before it was sold to Parkfield Castings, Horwich. It was later brought by the South Yorkshire Railway Preservation Society and moved to Meadowhall, later moving to Barrow Hill before heading to Scotland.

**No. 03099 (D2099), 3 September 2017**
No. 03099 stands among many other preserved shunters at the Heritage Shunter Trust depot, Rowsley. No. 03099 was withdrawn in 1976 from Thornaby depot and was sold to the National Coal Board, being used at both Fishburn and Monkton coking plants. It was initially preserved at Meadowhall before moving to Rowsley.

**No. 03112 (D2112), 14 August 2016**
D2112 is seen stabled at Robertsbridge on the Rother Valley Railway. This was withdrawn in 1987 and was initially used at the Nene Valley Railway. It was later used to shunt at Boston Docks before moving to Robertsbridge.

**No. 03113 (D2113), 20 April 2003**
No. 03113 stands in the Heritage Shunter Trust yard at Rowsley carrying Gulf Oil blue livery. This had been withdrawn from BR service in 1975, being sold directly to Gulf Oil for use at Milford Haven. When its working career had finished it was donated to the Heritage and Maritime Museum, Milford Haven, before arriving at Rowsley in 2003.

**No. 03113, 3 September 2017**

No. 03113 is seen preserved and fully restored to BR condition inside the Heritage Shunter Trust depot at Rowsley. This had arrived in 2003 carrying Gulf Oil Milford Haven blue livery.

**D2117, 18 June 2016**

D2117 stands outside the main shed at Haverthwaite on the Lakeside and Haverthwaite Railway carrying black livery. This was withdrawn in 1971 before it gained a TOPS number, and was sold directly to the L&HR.

**No. 03118 (D2118), 18 June 2006**

D2118 is seen in the Heritage Shunter Trust yard at Rowsley. Withdrawn in 1972, this was sold to Anglian Building Products for use at their site in Norfolk. It later moved to Dowmac of Tallington, Lincolnshire, in whose livery it is seen. Initially preserved at Meadowhall, it later moved to the Rutland Railway Museum before finally moving to Rowsley.

**No. 03118 (D2118), 28 December 2017**

No. 03118 is seen in the winter sunshine at Ruddington on the Great Central Railway (North). This had moved here from Rowsley and is seen having been fully restored and repainted into BR blue livery.

**No. 03120 (D2120), 28 May 2016**
D2120 is seen giving passenger rides at Didcot Railway Centre, coupled with former GWR Autocoach W231. D2120 was one of a handful of Class 03 shunters that received cut-down cabs while in BR service to work the Burry Port & Gwendreath Valley Line. It was withdrawn in 1986 and was sold directly to Sir William McAlpine, who moved it to his railway at Fawley Hill.

**No. 03128/03901 (D2128), 28 September 2019**
D2128 is seen stabled waiting its next turn of duty at Midsomer Norton station, which is the base for the Somerset & Dorset Railway Heritage Trust. This locomotive is privately owned and is currently in industrial use.

**No. 03901/03128 (D2128), 23 June 2018**
No. 03901 undergoes maintenance inside Briddon Engineering, Darley Dale. This loco has been renumbered into the 03/9 series as it is now fitted with a Cummins engine rather than a Gardner. This was withdrawn in 1976 carrying the number '03128' and was exported to Belgium to work for T. Zeebouw-Zeezand. It was repatriated in 1993 and has been in use at Scunthorpe Steel Works before moving to Darley Dale.

**D2133, 11 June 2017**
D2133 is seen looking resplendent at Minehead station on the West Somerset Railway. This was withdrawn in 1969 after only nine years' service and was sold to British Cellophane in Bridgewater. It was donated to the WSR in 1996 after its industrial use had finished.

**No. 03134 (D2134), 18 August 2018**
D2134 is seen stabled at the Royal Deeside Railway, Aberdeenshire, in front of the unique Battery Multiple Unit. D2134 was withdrawn in 1976 and, like its classmate D2128, was sold for use in Belgium with T. Zeebouw-Zeezand. It was also was repatriated, but in 1995 to Meadowhall, before moving to Scotland in 2001.

**D2138, 25 November 2017**
D2138 is seen as part of a large shunt move at the Midland Railway Centre, Butterley. Withdrawn in 1969, it was sold to the National Coal Board, being used at Bestwood Colliery and then Pye Hill No. 2 Colliery. It was preserved at Butterley in 1985.

**D2139, 3 September 2017**
D2139 stands outside the Heritage Shunter Trust shed at Rowsley having been restored to as-built condition. This was withdrawn as long ago as 1968, and was sold to the National Coal Board for use at various sites, including Gwent, Nantgarw, Coed Ely and Monkton Coking Plant. It was sold to the South Yorkshire Railway Preservation Society in 1992 and moved to Meadowhall, before moving finally to Rowsley.

**No. 03141 (D2141), 9 April 2017**
No. 03141 stands in Furnace Sidings on the Pontypool & Blaenavon Railway awaiting restoration. This was another fitted with a cut-down cab and was withdrawn from service in 1985. It was sold to two different scrap dealers before being resold for preservation.

**No. 03144 (D2144), 6 October 2018**
No. 03144 rests outside the shed at Leeming Bar on the Wensleydale Railway. This is also fitted with a cut-down cab from its BR days in South Wales. It was withdrawn in 1986 and was sold to the Territorial Army, who have based it at Wensleydale.

**No. 03145 (D2145), 14 May 2017**
No. 03145 is seen at Moreton-in-Marsh while giving brake van rides, along with Class 05 D2578. Owned by the Class D2578 Locomotive Group, this loco has a cut-down cab and was withdrawn in 1985, having a similar history to classmate No. 03141, before finding its way into preservation.

**D2148, 5 March 2016**
D2148 is seen on passenger duty at Preston Riverside station, part of the Ribble Steam Railway. This was withdrawn back in 1972 and was sold to the National Coal Board for use at Bowers Row Disposal Point. It was sold for preservation in 1987 and was moved to Steamport Southport. When Steamport closed in 1998 it was moved, along with the rest of the collection, to the Ribble Steam Railway.

**No. 03152 (D2152), 17 July 2016**
D2152 stands outside the shed at Hayes Knoll on the Swindon & Cricklade Railway. Fitted with a cut-down cab, this was withdrawn in 1983 and was sold directly into preservation at the S&CR.

**No. 03158 (D2158), 20 April 2003**
No. 03158 is seen in use at Peak Rail carrying black livery and the name *Margaret-Ann*. This was another Class 03 sold directly into preservation from BR service, being withdrawn in 1987. It has since moved to Mangapps Farm Railway Museum.

**No. 03162 (D2162), 22 May 1996**
D2162 is seen in as-withdrawn condition at Llangollen, carrying green livery. This was withdrawn in 1989 from Birkenhead depot and was sold directly into preservation.

**No. 03162 (D2162), 20 August 2017**
No. 03162 is seen stored at Pentrefelin sidings on the Llangollen Railway awaiting overhaul. By this time the loco had been repainted into BR blue livery, but shows signs of hard work.

**No. 03170 (D2170), 20 August 2016**
No. 03170 waits its next turn of duty at North Weald depot on the Epping & Ongar Railway. This was withdrawn in 1989 and was sold to Faber Prest in Salford for further use. It later came under the ownership of Otis Euro Transrail, who sold it on for preservation in 1999, with it being used at the Battlefield Line and later at Epping.

**D2178, 24 September 2017**

D2178 stands outside the main depot building at Bronwydd Arms on the Gwili Railway. This was another loco sold to the National Coal Board, being withdrawn in 1969. It was used at Gwaun-cae-Gurwen, Coed Ely and Nantgarw before being sold into preservation in 1985.

**No. 03179 (D2179, No. 97807), 23 September 2017**

No. 03179 is seen undergoing restoration at Rushden on the Rushden, Higham & Wellingborough Railway. This loco was used on the Isle of Wight during its BR career, being fitted with a cut down cab to work through the Ryde tunnels. It was later used at Horney depot for First Capital Connect, being the last Class 03 to be withdrawn from service in 2008. It remained at Hornsey until 2016, when it moved to Rushden.

**No. 03180 (D2180), 3 September 2017**
No. 03180 stands outside the Heritage Shunter Trust depot at Rowsley minus many parts, including windows and engine. This was withdrawn in 1984 and sold to Meyer Newman for use as a pilot at their Snailwell scrapyard. It was moved to Meadowhall in 1991 for preservation.

**D2182, 30 July 2017**
D2182 is seen in use as a depot pilot at Winchcombe carriage depot on the Gloucestershire & Warwickshire Railway. This was withdrawn from BR service in 1968, being sold to the National Coal Board and used at various sites including Glyn Neath, Gatewen, New Broughton, Wentworth and Hugglescote. It was then used in a children's playground in Leamington Spa before moving to the G&WR in 1986.

**D2184, 23 September 2017**
D2184 is seen at Castle Hedingham on the Colne Valley Railway carrying black livery. This was sold by BR in 1968 to the Coal Concentration depot at Southend for further use. It was preserved by the Colne Valley Railway in 1986.

**No. 03189 (D2189), 2 June 2019**
No. 03189 undergoes restoration inside the Furness Building at the Ribble Steam Railway. It was withdrawn in 1986 and sold for preservation to Steamport Southport, later moving to the Ribble Steam Railway in 1998 when Steamport closed.

**D2192, 18 July 2019**

D2192 is seen in use as a pilot at Churston on the Dartmouth Steam Railway. This loco carries black livery and the name *Titan*. It was withdrawn in 1970 and was sold directly into preservation, initially at the Dart Valley Railway before moving to Dartmouth.

**No. 03197 (D2197), 1 October 2017**

No. 03197 waits departure time from Smallbrook Junction on the Isle of Wight Steam Railway. It was visiting from Mangapps Farm Railway Museum for their diesel gala. It was withdrawn in 1987, initially for scrap to CF Booth, but it was later sold for preservation, being moved to Meadowhall.

**D2199, 11 August 2019**
D2199 is seen undergoing overhaul inside the Heritage Shunter Trust maintenance depot at Rowsley. This was withdrawn in 1972 and was sold to the National Coal Board for use at Rockingham Colliery. It was later sold for preservation and moved to Meadowhall. It has also seen industrial use with Eurotunnel at Dollands Moor when it was fitted with air brakes only.

**No. 03371 (No. 92, D2371), 18 July 2019**
No. 03371 is seen stabled at Churston on the Dartmouth Steam Railway. This was initially numbered Departmental 92 and was based at Cambridge, being renumbered D2371 in 1967. It was withdrawn in 1987 and was privately preserved before moving to Dartmouth.

**No. 03399 (D2399), 30 August 2006**
No. 03399 is seen stabled outside the shed at Mangapps Farm Railway Museum. This was the last Class 03 to be built in 1961 and was preserved at Mangapps directly from BR service in 1987.

# Class 04

**No. 11103 (D2203), 23 June 2019**
D2203 is seen outside the main shed building at Embsay at the Embsay & Bolton Abbey Railway. Over 140 Class 04 locomotives were built, and D2203 is the only surviving locomotive that was used on the Wisbech and Upwell Tramway. It was withdrawn in 1967 and was sold to the Hemelite Concrete Company, Herts, before being sold for preservation in 1982.

**No. 11106 (D2205), 3 September 2017**
D2205 looks superb outside the Heritage Shunter Trust depot at Rowsley, having been returned to BR green livery. It was withdrawn in 1969 and was sold to the Middlesbrough & Hartlepool Dock Authority, and later moved to Grangetown. It entered preservation in 1983 at the Kent & East Sussex Railway before moving to the West Somerset Railway and finally moving to Rowsley.

**No. 11108 (D2207), 20 May 2017**
D2207 is seen undergoing a very thorough restoration inside the depot at Grosmont on the North Yorkshire Moors Railway. Like classmate D2203, it was brought by Hemelite Concrete Company, Herts, before being sold to the NYMR for preservation in 1973.

**No. 11135 (D2229), 11 August 2019**
D2229 is seen at Rowsley at the start of a much-needed overhaul. Withdrawn in 1969, it passed to the National Coal Board for use at Brookhouse Colliery. It was also used at Orgreave and Manton Collieries before being sold into preservation at Meadowhall in 1990.

**No. 11215 (D2245), 23 August 2009**
D2245 stands outside the shed at Shackerstone on the Battlefield Line. This was withdrawn from BR service in 1968 and was sold direct into preservation, initially at the Derwent Valley Light Railway, before moving to Shackerstone in 1978.

**No. 11216 (D2246), 3 November 2017**
D2246 is seen in use as yard pilot at Buckfastleigh on the South Devon Railway. Withdrawn in 1968, it was sold to the National Coal Board for use at Crawley New Yard before moving to the Tolworth at West Drayton. It was preserved in 1994 at Meadowhall before being sold to the South Devon Railway in 2001.

**D2271, 11 June 2017**

D2271 is seen part way through an overhaul at Minehead on the West Somerset Railway. It was withdrawn in 1969 and was initially used by CF Booth as a yard shunter at its Rotherham scrapyard. It was preserved at the Midland Railway Centre in 1975 before moving to the West Somerset Railway in 1982. It can now be found at the South Devon Railway.

**D2272, 3 September 2017**

D2272 is seen awaiting restoration at the Heritage Shunter Trust, Rowsley. It was withdrawn from BR service in 1970 and was sold to British Fuels, Blackburn, before being preserved at Meadowhall in 1997.

**No. 11249 (D2279) 30 August 2006**
No. 11249 is seen standing in the yard at Chappel & Wakes Colne on the East Anglian Railway Museum. This loco only ever carried its 'D' prefix number in BR service, and was withdrawn in 1971. It was sold to the Central Electricity Generating Board for use at their Rye House Power Station. It was preserved by the EARM in 1981.

**D2284, 3 September 2017**
D2284 is seen fully restored outside the Heritage Shunter Trust depot at Rowsley. It was withdrawn in 1971 and was sold to the National Coal Board for use at North Gawber, but was also used at Grimethorpe and Woolley Colliery before entering preservation in 1985 at Meadowhall.

**D2289, 11 August 2019**

D2289 stands under tarpaulin at Rowsley. This loco was withdrawn in 1971 and was sold for further use to P. Wood, Queenborough, before being exported to Feralpi, Acciaierie di Lonato in Brescia, Italy. It was repatriated back to Rowsley in 2018.

**D2302, 14 May 2017**

D2302 is seen waiting overhaul at Moreton Park in Herefordshire. This was withdrawn in 1969 and was sold to the British Sugar Corporation for use at Woodston in Peterborough. It later moved to their Allscot factory in Wellington. It was later sold to G. G. Papworth in Ely before being preserved at Meadowhall in 1993.

**No. 04110 (D2310), 31 December 2018**

No. 04110 is seen stabled at Shackerstone on the Battlefield Line. It only ever carried its D2310 number in BR service, and was withdrawn in 1969, being sold to Coal Mechanisation Ltd, Tolworth, London. It was brought for preservation in 1994 and moved to Meadowhall.

**D2324, 9 July 2006**

D2324 stands among many other shunt engines at Barrow Hill awaiting restoration. It was withdrawn from York depot in 1968 and was sold to G. W. Talbot Ltd for use at their Aylesbury depot. It was later sold to Redland Roadstone in 1989, in whose livery it is seen, before entering preservation in 1995 at Meadowhall.

**D2325, 17 June 2017**
D2325 stands side by side with No. 03081 *Lucie* at Mangapps Farm Railway Museum. D2325 was withdrawn in 1968 and sold to the National Coal Board for use at their Norwich Coal Concentration depot. It finally entered preservation in 1986, moving to Mangapps Farm in 1988.

**D2334, 10 June 2017**
D2334 is seen in use as depot pilot at Dereham on the Mid-Norfolk Railway. It was withdrawn in 1968 and was brought by the National Coal Board for use at Manvers Main, but was also used at Thurcroft, Dinnington Main and Manton collieries. It was sold for preservation in 1988, moving to Meadowhall. It later moved to the Churnet Valley Railway before being brought by the Mid-Norfolk Railway.

**D2337, 27 June 1999**
D2337 is seen in preservation at Meadowhall, near Sheffield. Like many of its classmates, D2337 was withdrawn in 1968 and was sold to the National Coal Board for use at Manvers Main, but also saw use at Barnburgh Main Colliery. It was preserved in 1988 at Meadowhall. In this view it carried NCB livery.

**D2337, 3 September 2017**
D2337 is seen at Rowsley having been fully overhauled, and now carries BR green livery, replacing the NCB yellow livery.

# Class 05

**D2554 (Nos 11140, 05001, 97803), 1 October 2016**
D2554 stands outside the steam shed at Haven Street on the Isle of Wight Steam Railway. This was moved to Ryde St Johns depot in 1973 and was the only member of the class to receive a TOPS number. It was later renumbered '97803' when it entered departmental service, and was preserved in 1984. It carries the name *Nuclear Fred* and has been fitted with a cut-down cab.

**D2578, 14 May 2017**
D2578 stands in the sunshine at Moreton Park. This is one of sixty-nine shunters built by Hunslet from 1955 onwards – four survive in preservation. D2578 was withdrawn in 1967 and was sold to Bulmers of Hereford and preserved close to where it was used in private use.

**D2587, 11 August 2019**
D2587 stands outside the Heritage Shunter Trust depot at Rowsley. Withdrawn in 1967, it was sold to Chadderton Power Station, but was also later used at Kearsley Power Station. It entered preservation in 1983 at the East Lancashire Railway, before moving to the HST.

**D2595, 5 March 2016**
D2595 stands outside the shed at the Ribble Steam Railway. This led a very similar life to classmate D2587, being used at Chadderton Power Station and Kearsley Power Station before being sold to the East Lancashire Railway in 1983. It later moved to Steamport, Southport, and moved on to the Ribble Steam Railway when Steamport closed in 1998.

# Class 06

**No. 06003 (D2420, No. 97804), 9 July 2006**

No. 06003 is seen at Barrow Hill. This is the only surviving member of a class of thirty-five engines supplied by Andrew Barclay from 1958 onwards. No. 06003 survives by virtue of it being taken into departmental service in 1981 and was used at Reading carrying the number '97804'. It can currently be found at Rowsley.

# Class 07

**No. 07001 (D2985), 3 September 2017**
No. 07001 is seen giving brake van rides at Rowsley, being part of the Heritage Shunter Trust. It was one of fourteen shunters built in 1962 by Ruston, and all were based at Southampton Docks. No. 07001 was withdrawn in 1977 and was sold to the Staveley Lime Company in Buxton before being preserved at Meadowhall in 1989.

**No. 07010 (D2994), 9 April 2017**
No. 07010 is seen looking a little forlorn outside the shed at Bitton on the Avon Valley Railway. It was withdrawn in 1976 and was sold directly into preservation, initially at the Watercress Line, before moving to the West Somerset Railway in 1980. It moved to the Avon Valley Railway in 1994.

**No. 07012 (D2996), 25 November 2017**
No. 07012 stands on one of the roundhouse roads at Barrow Hill carrying BR blue livery. This was withdrawn in 1977 and was sold to P. D. Fuels Cwm Mawr Disposal Point before moving to Coed Bach Disposal Point. It was preserved at Meadowhall in 1988 and can be found today at Scunthorpe Steelworks giving rides around the works.

**No. 07013 (D2997), 17 November 2012**
No. 07013 is seen stabled outside Barrow Hill with classmate No. 07001 to the right. No. 07013 was withdrawn in 1977 and was sold to Dow Chemicals for use at their Kings Lynn depot. It was preserved at Meadowhall in 1994 and is today based at the East Lancashire Railway.

# Class 08

**No. 13000 (D3000), 3 September 2017**
No. 13000 stands inside the Heritage Shunter Trust shed at Rowsley. This was the very first Class 08 built out of almost a thousand examples. It was withdrawn in 1972 and sold to the National Coal Board for use at Hafodyrynys Colliery, but also found use at Bargoed, Mountain Ash and Mardy collieries. It was preserved in 1986 at the Preston Park Pullman sheds before moving to Meadowhall.

**No. 13002 (D3002), 18 November 2018**
No. 13002 is seen stabled in the platform at Plympton on the Plym Valley Railway. This was sold to Foster Yeoman for use at their Merehead Terminal in 1972 and was preserved at Plympton in 1982.

**No. 13014 (D3014), 18 July 2019**
D3014 looks resplendent in BR green livery at Paignton on the Dartmouth Steam Railway. Withdrawn in 1972, this was another shunter sold to the National Coal Board, this time for use at Merthyr Vale Colliery. It was preserved at Paignton in 1989.

**No. 13018/08011 (D3018), 4 May 2019**
D3018 is seen in use as yard pilot at Princes Risborough on the Chinnor & Princes Risborough Railway. This loco was sold directly into preservation when it was withdrawn from BR service in 1991 and carries the name *Haversham*.

**No. 13019 (D3019), 20 August 2017**
D3019 is seen part way through a restoration at Llynclys – part of the Cambrian Heritage Railway. Withdrawn in 1973, it was sold to P. D. Fuels LTS Gwaun-cae-Gurwen and was preserved at Meadowhall in 2000.

**No. 13022/08015 (D3022), 21 May 2017**
D3022 shunts GWR carriages into the platform at Kidderminster on the Severn Valley Railway. This loco was brought directly from British Rail by the SVR in 1983.

**No. 08016 (No. 13023, D3023), 3 September 2017**
No. 08016 is seen carrying BR blue livery outside the Heritage Shunter Trust shed at Rowsley. This loco was withdrawn in 1980 and was sold to Hargreaves LTD for use at British Oak Coal Disposal Point. It was preserved in 1989, initially at Meadowhall before moving to Rowsley.

**No. 08022 (No. 13030, D3030), 30 May 2016**
No. 08022 *Lion* is seen after having arrived at Cholsey on a passenger turn on the Cholsey & Wallingford Railway. Withdrawn in 1985, it was sold to Guinness for use at their Park Royal depot and has been on loan at Cholsey since 1997.

**No. 08032 (No. 13044, D3044), 6 October 2019**
No. 08032 is seen among the undergrowth at Ropley on the Mid-Hants Railway awaiting overhaul. This was withdrawn in 1974 and was sold to Foster Yeoman for use at their Merehead site, gaining the number '33'. It moved to Ropley in 2008.

**No. 3059 (No. 13059, D3059, No. 08046), 18 August 2018**
No. 3059 is seen in use on brake van rides at Bridge of Dun on the Caledonian Railway. Withdrawn in 1980, it was sold to Associated British Maltsters for use at Airdrie. It was preserved at the Caledonian Railway in 1985 and has been named *Brechin City*.

**No. 08054 (No. 13067, D3067), 23 June 2019**
No. 08054 rests at Bolton Abbey on the Embsay & Bolton Abbey Railway. This was withdrawn from BR service in 1980 and was sold to Tilling Construction for use at their Grassington Works in North Yorkshire. It moved to Embsay in 2008.

**No. 08060 (No. 13074, D3074), 30 May 2016**
No. 08060 *Unicorn* is seen at Cholsey on the Cholsey & Wallingford Railway. This locomotive's history mirrors that of classmate No. 08022, and it too has been on loan at Cholsey from Guinness Park Royal since 1997.

**No. 13079 (D3079, No. 08064), 1 September 2019**
No. 13079 is seen stabled inside the depot at the National Railway Museum, York. Withdrawn in 1985, it was claimed by the NRM straight from British Rail.

**No. 13101 (D3101), 28 December 2018**
No. 13101 is seen stabled among steam engines at Loughborough on the Great Central Railway. This was withdrawn in 1972 and was sold to ARC (East Midlands) and moved to their Loughborough site. It was preserved at the Great Central Railway in 1984.

**No. 13180 (D3180, No. 08114), 28 December 2017**

No. 13180 is seen in the yard at Ruddington on the Great Central Railway (North). It is seen part way through an overhaul, which would result in it being painted into BR blue and numbered '08114'. This was withdrawn in 1983 and was initially preserved at the Gloucestershire & Warwickshire Railway before moving to the Swindon & Cricklade Railway. It arrived at Loughborough in 1991.

**D3201 (No. 13201/08133), 22 May 2016**

D3201 is seen undergoing a very thorough overhaul inside the diesel repair depot at Kidderminster on the Severn Valley Railway. This was withdrawn in 1980 and sold to the Sheerness Steel Company for use at their Isle of Sheppey site. It was sold into preservation at Meadowhall in 1995 and arrived at Kidderminster in 2003.

**No. 08168 (No. 13236, D3236), 11 June 2006**
No. 08168 is seen stabled at Shackerstone on the Battlefield Line. No. 08168 was withdrawn from BR service in 1988 and was sold to BREL for use at York works. It has since spent time at various industrial locations. It is privately owned and is currently in use at Burton depot.

**No. 13261 (D3261), 17 July 2016**
D3261 is seen inside the shed at Hayes Knoll on the Swindon & Cricklade Railway undergoing repairs. This was withdrawn in 1972 and was sold to the National Coal Board for use at Tower Colliery. It was sold for preservation in 1987, and moved to Preston Park Pullman shed, before moving to Swindon in 1997.

**No. 13265 (D3265, No. 08195) 20 August 2017**

No. 13265 is seen at the head of an engineers' train at Llangollen shed. Withdrawn in 1983, it was sold directly to the Llangollen Railway, arriving in 1986.

**No. 08202 (No. 13272, D3272), 28 September 2019**

No. 08202 is seen stabled at Bitton on the Avon Valley Railway. This loco was withdrawn in 1989 and was sold to G. G. Papworth, which later became the Potter Group, for use at their Ely site. This was preserved in 2014, initially at the Chasewater Railway before moving to the Avon Valley.

**No. 08238 (No. 13308, D3308), 17 July 2019**

No. 08238 *Charlie* is seen stabled around the back of the signal box at Lydney Junction on the Dean Forest Railway. Withdrawn in 1984, it was initially sold to Forest Free Mining, but the sale was never completed. It was later purchased by Swindon Railway Workshop and moved there in 1988. It was later moved to Lydney in 1993.

**No. 08266 (No. 13336, D3336), 22 August 2016**

No. 08266 is seen in the depot yard at Haworth on the Keighley & Worth Valley Railway, looking resplendent in departmental grey livery. Withdrawn in 1985, this was sold directly to the K&WVR for preservation.

**No. 08288 (D3358), 6 October 2019**
D3358 stands outside the depot at Ropley on the Mid-Hants Railway. This was withdrawn from BR service in 1983 and was sold directly to the Mid-Hants Railway, arriving in 1984.

**No. 08331 (D3401), 28 May 2017**
No. 08331 is seen stabled inside the West Shed at the Midland Railway Centre, Butterley. It was withdrawn in 1988 and was sold to RFS Engineering until it was sold into preservation in 2007 – initially at the Embsay & Bolton Abbey Railway and has been at Butterley since 2008.

**No. 08359 (D3429), 22 April 2018**
D3429 is seen stabled at Brownhills West on the Chasewater Railway. Withdrawn from BR service in 1984, D3429 was sold to the Churnet Valley Railway. It would later move to Tyseley, then later to the Chasewater Railway.

**No. 08377 (D3462), 6 October 2019**
No. 08377 is seen as yard pilot at Arlesford on the Mid-Hants Railway, carrying BR green livery but with a TOPS number. Withdrawn in 1983, it was initially sold to the Dean Forest Railway before moving to the West Somerset Railway in 1996. It was eventually sold to the Mid-Hants Railway in 2013.

**No. 08436 (D3551), 12 August 2017**
No. 08436 is seen at Swanage carrying London & South Western Railway livery. This was withdrawn in 1992 and initially moved to Meadowhall before being sold to RT Railtours in 1998. It arrived at Swanage in 2004.

**No. 08443 (D3558), 16 August 2018**
D3558 sits in the depot yard at Bo'ness on the Bo'ness & Kinneil Railway. This was withdrawn in 1985 and was sold to DCL Scottish Grain Distillers for use at their Alloa site. It was preserved at Bo'ness in 1995.

**No. 08471 (D3586), 19 May 2019**
D3586 is seen on depot pilot duty at Bridgnorth on the Severn Valley Railway. This was withdrawn in 1985 and was sold directly to the SVR for preservation.

**No. 08476 (D3591), 12 August 2017**
D3591 is seen coupled to a BR Mk I carriage at Swanage. D3591 was withdrawn in 1985 and was sold directly to the Swanage Railway for preservation.

**No. 08490 (D3605), 17 August 2018**
D3605 stands in the depot yard at Aviemore on the Strathspey Railway alongside Class 31 D5862. D3605 was withdrawn in 1985 and was another Class 08 sold directly into preservation, arriving at Aviemore in 1987.

**No. 08495 (D3610), 20 May 2017**
No. 08495 is seen inside the running shed at Grosmont on the North Yorkshire Moors Railway carrying EWS livery and remote-control equipment. This loco didn't enter preservation until 2017.

**No. 08528 (D3690), 28 December 2018**
D3690 is seen stabled at Loughborough on the Great Central Railway. This was withdrawn from EWS service in 2010, initially moving to the Battlefield Line before moving to Loughborough in 2014.

**No. 08556 (D3723), 20 May 2017**
No. 08556 is seen giving brake van rides at Pickering on the North Yorkshire Moors Railway. This loco was withdrawn from BR service in 1990 and was sold directly to the North Yorkshire Moors for preservation, arriving in 1993.

**No. 08590 (D3757), 3 September 2017**
No. 08590 is seen acting as yard pilot at Swanwick at the Midland Railway Centre. Withdrawn in 1993, it moved directly to the Midland Railway Centre for preservation and carries the name *Red Lion*.

**No. 08993 (D3759, No. 08592), 22 August 2016**
No. 08993 *Ashburnham* is seen carrying EWS maroon livery at Haworth on the Keighley & Worth Valley Railway. This was one of five Class 08s that were fitted with cut-down cabs for use on the Burry Port & Gwendreath Valley line. It was preserved at the K&WVR in 2016.

**No. 08604 (D3771), 24 February 2002**

No. 08604 *Phantom* is seen outside the steam shed at the Great Western Railway society in Didcot. Withdrawn in 1993, this loco was sold directly into preservation, arriving at Didcot in 1994.

**No. 08605 (D3772), 11 August 2019**

No. 08605 is seen in the depot yard at Ecclesbourne. This loco is seen carrying remote-control equipment and the name *G. R. Walker*. It entered preservation at the Ecclesbourne Railway in 2018.

**No. 08635 (D3802), 21 May 2017**
No. 08635 is seen inside the diesel depot at Kidderminster on the Severn Valley Railway. This arrived on the line in 2007 for preservation, but will be used as a test bed for hydrogen-battery power.

**No. 08757 (D3925), 17 September 2017**
No. 08757 is seen inside the small shed at Telford on the Telford Steam Railway. This loco was withdrawn from EWS service in 2013 and was stored at Crewe Electric depot, before being sold in 2017 for preservation at Telford.

**No. 08767 (D3935), 29 August 2018**
D3935 is seen on station pilot duty at Sheringham on the North Norfolk Railway, having been beautifully restored carrying BR green livery. Withdrawn in 1994, it was sold directly to the NNR.

**No. 08769 (D3937), 17 July 2019**
D3937 *Gladys* is seen at Lydney Junction on the Dean Forest Railway. Withdrawn in 1990, this loco was sold to the Fire Service Training College at Moreton-in-Marsh. It was sold to the Dean Forest Railway for preservation and arrived in 2000.

**No. 08773 (D3941), 23 June 2019**
D3941 stands outside the carriage shed at Embsay on the Embsay & Bolton Abbey Railway. Withdrawn in 1998, this loco was sold to a private individual, who moved the loco to Embsay in 2006.

**No. 08830 (D3998), 3 September 2017**
No. 08830 is seen in the yard at Peak Rail carrying BR blue livery. This loco was withdrawn from BR service in 1996 and was used by Prism Rail, who loaned the loco to the East Somerset Railway. It has been used at various industrial sites, including Mendip Rail. It was sold to Waterman Railway Heritage Trust and has been at Peak Rail since 2015.

**No. 08850 (D4018), 20 May 2017**
No. 08850 is seen on one of the turntable roads at Pickering on the North Yorkshire Moors Railway. Withdrawn in 1992, this was sold to the West Somerset Railway before moving to Yorkshire in 1998.

**No. 08881 (D4095), 28 September 2019**
D4095 is seen in the platform at Midsomer Norton on the Somerset & Dorset Railway Heritage Trust. This was withdrawn from EWS service in 2004 and was initially preserved at the Gloucestershire & Warwickshire Railway before moving to Midsomer Norton in 2011.

**No. 08907 (D4137), 27 May 2019**
D4137 is seen at Loughborough station on the Great Central Railway while on station pilot duty. This was withdrawn from DB Schenker service in 2014 and arrived at the GCR for preservation in 2016.

**No. 08911 (D4141), 7 October 2017**
No. 08911 *Matey* is seen outside Locomotion Shildon while on yard pilot duty. This loco was withdrawn in 2004 and was sold directly to the National Railway Museum as a working exhibit.

**No. 08944 (D4174), 16 April 2016**
No. 08944 is seen at the East Lancashire Railway while receiving attention. This loco was withdrawn from service in 1998 and was sold to RFS Engineering for industrial use. It later moved to the East Lancashire Railway in 2000.

# Class 09

**No. 09001 (D3665), 3 September 2017**
No. 09001 is seen inside the Heritage Shunter Trust shed at Rowsley. This Class 09 locomotive was withdrawn from service in 2010 and has been beautifully restored to working order, carrying EWS maroon livery.

**No. 09010 (D3721), 3 November 2017**
D3721 is seen in the yard at Buckfastleigh on the South Devon Railway. This loco was withdrawn from service in 2004 and was later sold directly to the South Devon Railway in 2010.

**No. 09012 (D4100), 19 May 2019**
D4100 *Dick Hardy* is seen outside the diesel shed at Kidderminster on the Severn Valley Railway. It was withdrawn from DB Schenker service in 2010 and was initially sold to Harry Needle, before it was purchased by the SVR in 2013.

**No. 09015 (D4103), 28 September 2019**
No. 09015 is seen undergoing restoration at Bitton on the Avon Valley Railway. This loco was withdrawn from EWS service in 2008 and was sold to the National Railway Museum, but was later sold to Rail Support Services as a source of spare parts after No. 09017 came up for sale and was in a better condition. It was sold in 2019 to a private individual who moved the loco to Bitton.

**No. 09017 (D4105, No. 97806), 1 September 2019**
No. 09017 stands outside the National Railway Museum, York. This loco was withdrawn in 1987 and transferred to the departmental sector for use as the Severn Tunnel emergency loco, being numbered '97806'. It later reverted to No. 09017 and was finally withdrawn from EWS service in 2009, being brought by the NRM in 2011.

**No. 09018 (D4106), 26 October 2017**
D4106 is seen in use as a depot pilot at Sheffield Park on the Bluebell Railway. Withdrawn in 2006, this loco was sold to Harry Needle and was put into industrial use, including being used at the old Dalby test track. It was sold into preservation, arriving at the Bluebell Railway in 2013.

**No. 09019 (D4107), 11 June 2017**
D4107 is seen at Bishops Lydyard on the West Somerset Railway. Withdrawn in 2009, this loco was initially sold to Harry Needle before it was preserved at the West Somerset Railway in 2013.

**No. 09024 (D4112), 16 April 2016**
No. 09024 is seen at Heywood on the East Lancashire Railway during a shunter-running day. This was withdrawn from EWS service in 2009 and was initially moved to Moveright International at Wishaw, before being sold to the East Lancashire Railway in 2014.

**No. 09026 (D4114), 21 July 2016**
No. 09026 *Cedric Wares* is seen in the yard at Tunbridge Wells West on the Spa Valley Railway. This had only recently been preserved, being withdrawn from use with Southern Railway in 2016.

# Class 10

**D3452, 9 July 2016**
D3452 is seen stabled at Bodmin General on the Bodmin & Wenford Railway. There were nearly 150 Class 10 locomotives built, but all were withdrawn early as being non-standard. D3452 was withdrawn in 1968 and was sold to English China Clay for use at Fowey. It was preserved at Bodmin in 1989.

**D3476, 27 June 1999**
D3476 is seen preserved at Meadowhall. This had been withdrawn back in 1968 and was brought by English China Clay for use at their Fowey works. It is seen still carrying its original BR livery but would be sold for scrap in 2009.

**No. 10119 (D4067), 27 May 2016**

No. 10119 stands outside the main depot at Loughborough on the Great Central Railway. This loco, like all the other Class 10 locomotives, never received a TOPS number in BR service. It was withdrawn in 1970 and was sold to the National Coal Board for use at Betteshanger Colliery, but was also used at Snowdown and Nailstone collieries. It was preserved at Loughborough in 1980 and now carries the name *Margaret Ethel – Thomas Alfred Naylor*.

**D4092, 25 November 2017**

D4092 stands inside the roundhouse at Barrow Hill, minus its coupling rods. This was withdrawn from BR service in 1986 and was sold to P. D. Fuels for use at their Gwaun-cae-Gurwen works. It was saved for preservation in 1988 at Meadowhall before moving to Barrow Hill.

# Class 11

**No. 12049 (Nos 12082 and 01553), 7 May 2016**
No. 12049 is seen in the yard at Ropley on the Mid Hants Railway. This is the second Class 11 locomotive to carry this number. The original No. 12049 was damaged beyond repair at Ropley in 2010 and was scrapped, with classmate No. 12082 assuming its identity.

**No. 12077, 21 November 2015**
No. 12077 stands in the sunlight in the depot yard at the Midland Railway Centre, Butterley. This was withdrawn from service in 1971 and was sold to J. Cashmore for use at their Great Bridge scrapyard. It was preserved at Butterley in 1978.

**No. 12082 (Nos 01553 and 12049), 9 June 2007**

No. 12082 is seen at MoD Long Marston carrying Harry Needle orange livery and running number '01553'. This was one of 120 Class 11 locomotives built, and like the Class 10 they were withdrawn early as non-standard. No. 12082 was withdrawn in 1971 and was sold to Shellstar Fertilisers Ltd for use at Ince. It was saved for preservation in 1991 at Meadowhall, but was loaned out to various industrial operators. It was renumbered '12049' in 2010 to take the place of its classmate at the Mid-Hants Railway.

**No. 12083 (No. 01563), 31 December 2018**

No. 12083 is seen in derelict condition at Market Bosworth on the Battlefield Line. This was withdrawn in 1971 and was sold to Tilcon for use at their Grassington Works in North Yorkshire. It was preserved in 1995 and moved to Meadowhall before moving to the Midlands. It still carries Tilcon livery.

**No. 12088 (No. 01564), 8 October 2017**
No. 12088 is seen undergoing an overhaul in the repair shed at Alnwick on the Aln Valley Railway. This was withdrawn in 1971 and was sold to the National Coal Board for use at their Swalwell Disposal Point. It was preserved at Meadowhall in 1989 but was loaned out to Widdrington Disposal Point in 1996, gaining the number '01564' in the process.

**No. 12093, 18 August 2018**
No. 12093 rests in the station yard at Brechin on the Caledonian Railway. Withdrawn from BR service in 1971, it was sold to the National Coal Board for use at Widdrington Disposal Point. It was preserved in 1987 at the Ayrshire Railway Preservation Society, but moved to Brechin in 2002.

**No. 12099, 22 May 2016**
No. 12099 is seen on station pilot duty at Kidderminster on the Severn Valley Railway. This was withdrawn in 1971 and sold to the National Coal Board for use at their Lion Disposal Point in Blaenavon. It moved around a great many different NCB sites before being sold for scrap to CF Booth in 1989. It was then sold to the Severn Valley Railway in 1990.

**No. 12131, 29 August 2018**
No. 12131 is seen in use as depot pilot at Weybourne on the North Norfolk Railway. Withdrawn in 1969, it was sold to the National Coal Board for use at Betteshanger Colliery in Kent and later at Snowdown Colliery. It arrived at the North Norfolk Railway for preservation in 1982.

# Class 12

**No. 15224, 21 July 2016**

No. 15224 is the only surviving member of the twenty-six-strong Class 12. Withdrawn in 1971, it was sold to the National Coal Board for use at Betteshanger Colliery before moving to Snowdown Colliery. It was preserved in 1982 and has been based at Preston Park Pullman Works and the Lavender Line before moving to Tunbridge Wells for use at the Spa Valley Railway where it is seen.

# Class 14

D9500, 23 June 2018
D9500 is seen waiting overhaul at Briddon Engineering, Darley Dale. This was the first of fifty-six Class 14 locomotives built by Swindon Works from 1964 onwards. D9500 was withdrawn in 1969. It was one of many sold to the National Coal Board for use at their Ashington Colliery. It was initially saved for preservation in 1987 at the Llangollen Railway, before heading to Meadowhall.

D9513, 23 June 2019
D9513 stands in the depot yard at Embsay on the Embsay & Bolton Abbey Railway. This was withdrawn in 1968 and was sold to the National Coal Board for use at their British Oak site. It was used at many NCB sites, including Burradon Colliery where it gained the number '38'. It moved to Embsay in 1987 and has been restored, carrying one of its NCB numbers and liveries.

**D9516, 28 May 2016**
D9516 is seen on display at Didcot Railway Centre. Withdrawn in 1968, it was sold to Stewart & Lloyd for use at Corby Quarry. It was eventually saved for preservation at the Great Central Railway in 1981 and was later moved to the Nene Valley Railway.

**D9520, 1 September 2016**
D9520 is seen in the depot yard at Wansford on the Nene Valley Railway along with Class 55 No. 55022 masquerading as No. 55007 *Pinza*. D9520 was withdrawn in 1968 and was sold to British Steel for use at Glendon East Quarry. It later moved to BSC Corby where it was numbered '45'. Preserved in 1981, it has been based at many different sites, but is now at the Nene Valley Railway.

**D9521, 30 July 2016**
D9521 leads classmate D9555 over the level crossing into Lydney Junction station on the Dean Forest Railway. This was withdrawn in 1969 and was sold to the National Coal Board for use at Ashington Colliery. It was preserved in 1987 at the Rutland Railway Museum, but can now be found at the Dean Forest Railway.

**D9523, 6 October 2018**

D9523 is seen carrying maroon livery at Leeming Bar on the Wensleydale Railway. This was another Class 14 that found use at British Steel Glendon East Quarry before moving to Corby. It was preserved in 1981 at the Great Central Railway before moving to the Nene Valley Railway. It later moved to the Wensleydale Railway in North Yorkshire.

**No. 14901 (D9524), 23 September 2017**

No. 14901 is seen at Castle Hedingham on the Colne Valley Railway carrying a fictitious BR TOPS number. This was withdrawn in 1969 and was sold to British Petroleum for use in Grangemouth. It later had its Paxman engine removed and a Dormer power unit was fitted. It was sold for preservation in 1981 and moved to the Bo'ness & Kinneil Railway before moving south to the Colne Valley Railway.

**D9525, 29 May 2016**
D9525 is seen undergoing overhaul inside the Heritage Shunter Trust shed at Rowsley. This was withdrawn in 1968 and was sold to the National Coal Board for use at their Philadelphia Colliery. It also operated at Ashington Colliery before being preserved at the Kent & East Sussex Railway in 1987. It has moved around many different preservation sites before moving to Rowsley.

**D9526, 11 June 2017**
D9526 is seen undergoing the final stages of an overhaul at Willington on the West Somerset Railway. This was withdrawn in 1969 and was sold to the Associated Portland Cement Company for use at their Westbury site. It was preserved at the WSR in 1980.

**No. 9529 (D9529), 1 September 2016**
No. 9529 stands outside the shed at Wansford on the Nene Valley Railway. This was another Class 14 sold to British Steel, but was used at Buckminster Ironstone Quarry. It later joined many of its classmates at Corby. It was preserved in 1980 at the North Yorkshire Moors Railway before eventually moving to the Kent & East Sussex Railway. It was loaned out to Victa Railfreight for use at Beechbrook Farm. It can now be found at the Nene Valley Railway.

**D9531, 16 April 2016**
D9531 arrives at Summerseat on the East Lancashire Railway. This was another Class 14 to find employment with the National Coal Board, eventually being used at Ashington Colliery. It was preserved by the ELR in 1987 and carries the name *Ernest*.

**D9537, 8 July 2016**
D9537 arrives at Bury Bolton Street on the East Lancashire Railway carrying desert sand livery. This was withdrawn in 1968 and was sold to British Steel for use at Gretton Brook, Corby. Initially it was preserved at the Gloucestershire & Warwickshire Railway before becoming resident at the Ecclesbourne Railway.

**D9537, 11 August 2019**
D9537 is seen at Wirksworth station on the Ecclesbourne Railway. In this view it has gained BR black livery.

**D9539, 5 March 2016**
D9539 is seen at Preston Riverside station on the Ribble Steam Railway. The 6E 32 headcode is used when the railway has oil trains working in from the main line. D9539 was withdrawn in 1968 and was sold to British Steel for use at Gretton Brook Quarry. It was initially preserved at the Gloucestershire & Warwickshire Railway in 1983 but has since moved to the Ribble Steam Railway.

**D9551, 19 May 2019**
D9551 stands outside Bridgnorth shed on the Severn Valley Railway carrying Golden Ochre livery. This was another Class 14 sold to British Steel for use in Corby in 1968. This was preserved in 1981 at the West Somerset Railway before moving to the Royal Deeside Railway in Aberdeenshire. It later moved to the Severn Valley Railway.

**D9553, 30 July 2006**
D9553 stands outside the depot at Toddington on the Gloucestershire & Warwickshire Railway. This is yet another Class 14 that was sold to British Steel Corby in 1968, where it gained the number '54'. It was preserved in 1983 by the G&WR but has since moved to the Vale of Berkeley Railway.

**D9555, 30 July 2016**
D9555 is seen at Lydney Junction on the Dean Forest Railway. This was withdrawn in 1969 and sold to the National Coal Board for use at their Burradon Colliery. It later moved to join its other classmates at Ashington Colliery. Preserved in 1987, it initially moved to the Rutland Railway Museum before finding its way to the Dean Forest Railway.

# Miscellaneous Shunters used by British Rail

**D2511, 22 August 2016**

D2511 is seen in use as depot pilot at Oxenhope on the Keighley & Worth Valley Railway. This was one of twenty locomotives built by Hudswell Clarke, with this one dating from 1961. It was withdrawn from service before the class could be given a TOPS classification. Sold in 1968 to the National Coal Board for use at Brodsworth Colliery, it would be preserved at the K&WVR in 1977.

**D2767, 16 August 2018**

D2767 is seen stabled outside the shed at Bo'ness on the Bo'ness & Kinneil Railway. This is one of eighty-one such locomotives built by North British from 1953 onwards. It was withdrawn in 1967 and sold to Burmah Oil Company for use at Stanlow. It was initially preserved at the East Lancashire Railway in 1983 but moved to Scotland in 2001.

**D2774, 17 August 2018**

D2774 stands inside the shed at Aviemore on the Strathspey Railway. This was withdrawn in 1967 and sold to the National Coal Board for use at their Killoch Colliery. It later moved to Celynen Colliery in Gwent before being preserved at the East Lancashire Railway in 1987. It moved north to Scotland in 2001.

**D0226, 22 August 2016**

D0226 is seen outside the shed at Haworth on the Keighley & Worth Valley Railway. This was one of a pair of demonstrator locomotives built by English Electric in 1956. They were both used at Stratford depot in London before being returned to English Electric in 1966. D0226 was preserved at the K&WVR and carries the name *Vulcan*.

**No. 10, 14 September 2019**
No. 10 is seen stabled outside the depot at Aberystwyth on the Vale of Rheidol Railway. This was built in 1987 using parts from Baguley-Drewry Engineering and is used to compliment the line's three narrow-gauge steam locomotives.

**ED10, 28 July 2019**
ED10 is seen preserved at the Irchester Narrow Gauge Railway. This narrow-gauge shunter was built by Ruston Hornsby in 1958 and was used by British Railways at Beeston Creosote/Sleeper factory until withdrawal in 1965. It carries the name *Edward Charles Hampton*.

**North British No 27414-1954, 17 September 2017**

North British 27414-1954 is seen preserved at the Telford Steam Railway carrying BR black livery and the name *Tom*. This was built as a demonstrator locomotive by North British and was used by BR at various Western Region places, including Highbridge, St Phillips Marsh and Danygraig. It was returned to North British in 1955 and was sold to GKN at Hadley. It was preserved at Telford in 1983.

**North British 27415-1954, 16 August 2018**

North British 27415-1954 is seen inside the museum at the Bo'ness & Kinneil Railway. This was another demonstrator built by North British and was used in London, being based at Devon's Road, Bow, and also West India Docks and Willesden. It later went to the Southern Region for trials at Southampton Docks. It returned to North British and was sold to Esso for use at Fawley before being preserved at the Isle of Wight Steam Railway in 1979. It moved to Scotland in 1991.

**PWM650 (No. 97650), 3 September 2017**
PWM650 is seen in preservation at the Heritage Shunter Trust, Rowsley. This loco was built by Ruston Hornsby in 1953 and was allocated the departmental number 'PWM650'. Also, like the other four members of the class, it was based on the Western Region. It was withdrawn in 1987 and plinthed outside Lincoln Steam Shed before it was moved to the Lincolnshire Wolds Railway. It moved to Rowsley in 2017.

**PWM651 (No. 97651), 17 July 2016**
PWM651 is seen inside the shed at Hayes Knoll on the Swindon & Cricklade Railway. This was built in 1959 and was used at various Western Region depots, but was later to be found at Gloucester Horton Road and Cardiff Canton. It was withdrawn in 1996 and was initially preserved at the Northampton & Lamport Railway before moving to the Strathspey Railway. It can now be found at Swindon.

**PWM654 (No. 97654), 3 September 2017**
PWM654 is seen giving brake van rides at Rowsley. This locomotive was withdrawn from service in 2005 and was brought by the Heritage Shunter Trust, who have restored it to BR blue livery.

**ZM32, 11 August 2019**
ZM32 stands at Steeple Grange on the Steeple Grange Light Railway. This narrow-gauge shunter was built by Ruston & Hornsby in 1957 for use at Horwich Works. It was preserved by the SGLR in 1997 and carries BR lined green livery, complete with *Horwich* nameplates.